by Spencer Brinker

Consultant:
Beth Gambro
Reading Specialist
Yorkville, Illinois

Contents

BEARPORT
PUBLISHING

New York, New York

A Rainy Day

Look!
It is rainy.

Today is rainy.

I see a gray cloud.

Today is rainy.

I see a red umbrella.

Today is rainy.

I see a white dog.

Today is rainy.

I see a green boat.

Today is rainy.

I see yellow boots.

Look around!

Teaching Tips

Before Reading

✔ Guide readers on a "picture walk" through the text by asking them to name the things shown.

✔ Discuss book structure by showing children where text will appear consistently on pages.

✔ Highlight the supportive pattern of the book. Note the consistent number of sentences and words found on each alternating page.

During Reading

✔ Encourage children to "read with your finger" and point to each word as it is read. Stop periodically to ask readers to point to a specific word in the text.

✔ Reading strategies: When encountering unknown words, prompt readers with encouraging cues, such as:

- **Does that word look like a word you already know?**
- **It could be _____ , but look at _____ .**
- **Check the picture.**

After Reading

✔ Write the key words on index cards.

- **Have readers match them to pictures in the book.**
- **Have children sort words by category (words that start with _b_ or _c_, for example).**

✔ Encourage readers to talk about different types of weather.

✔ Ask readers to identify their favorite page in the book. Have them read that page aloud.

✔ Ask children to write their own sentences about the weather. Encourage them to use the same pattern found in the book as a model for their writing.

Credits: Cover, © FamVeld/iStock; 2–3, © miller3181/iStock and © phototropic/iStock; 4–5, © ShaunWilkinson/Shutterstock; 6–7, © Dinodia Photos/Alamy; 8–9, © Ralf Bitzer/Alamy; 10–11, © triloks/iStock; 12–13, © evgenyatamanenko/iStock; 14–15, © stevecoleimages/iStock; 16T (L to R), © triloks/iStock and © evgenyatamanenko/iStock; 16B (L to R), © ShaunWilkinson/Shutterstock, © Ralf Bitzer/Alamy, and © Dinodia Photos/Alamy; Back Cover, © miller3181/iStock.

Publisher: Kenn Goin   **Senior Editor**: Joyce Tavolacci   **Creative Director**: Spencer Brinker

Library of Congress Cataloging-in-Publication Data in process at the time of publication (2019)
Library of Congress Control Number: 2018014543
ISBN-13: 978-1-68402-997-6 / ISBN: 978-1-64280-135-4 (paperback)

What can you see
on a rainy day?

15

Key Words

boat

boots

cloud

dog

umbrella

Index

About the Author

Spencer Brinker lives and works in New York City. Weather never stops him from enjoying the city.